VELOCIRAPTOR

Many dinosaurs once roamed Earth.

They lived long before people.

Some dinosaurs became **fossils** after they died.

Paleontologists study fossils to learn about dinosaurs.

WHEN DINOSAURS RULED THE EARTH

THE VELOCIRAPTOR

Written by Tracy Vonder Brink Illustrated by Riley Stark

TABLE OF CONTENTS

A Crabtree Seedlings Book

Crabtree Publishing
crabtreebooks.com

School-to-Home Support for Caregivers and Teachers

This book helps children grow by letting them practice reading. Here are a few guiding questions to help the reader with building his or her comprehension skills. Possible answers appear here in red.

Before Reading:

- What do I think this book is about?
  - *I think this book is about dinosaurs.*
  - *I think this book is about the dinosaur called* Velociraptor.
- What do I want to learn about this topic?
  - *I want to learn how big* Velociraptor *was.*
  - *I want to learn what* Velociraptor *ate.*

During Reading:

- I wonder why…
  - *I wonder why paleontologists study fossils.*
  - *I wonder how* Velociraptor *used its claws.*
- What have I learned so far?
  - *I have learned that dinosaurs lived before people.*
  - *I have learned that* Velociraptor *lived in Asia.*

After Reading:

- What details did I learn about this topic?
  - *I have learned that* Velociraptor *had three toes on each foot.*
  - *I have learned that* Velociraptor *ran faster than other dinosaurs.*
- Read the book again and look for the glossary words.
  - *I see the word* ***paleontologists*** *on page 4 and the word* ***carnivore*** *on page 16. The other glossary words are found on page 22.*

Velociraptor was a dinosaur that lived about 80 million years ago.

Velociraptor fossils have been found in Asia.

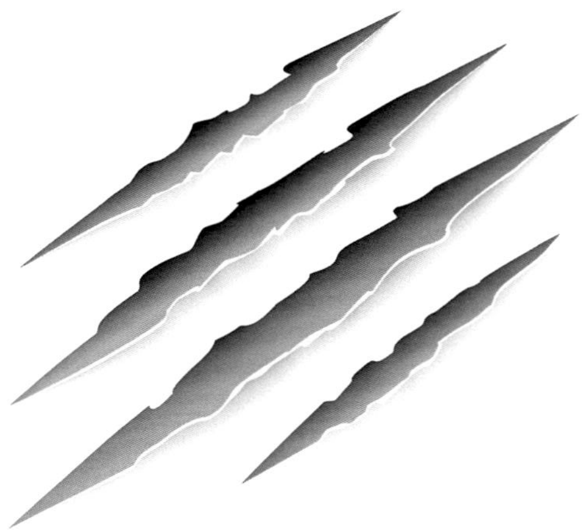

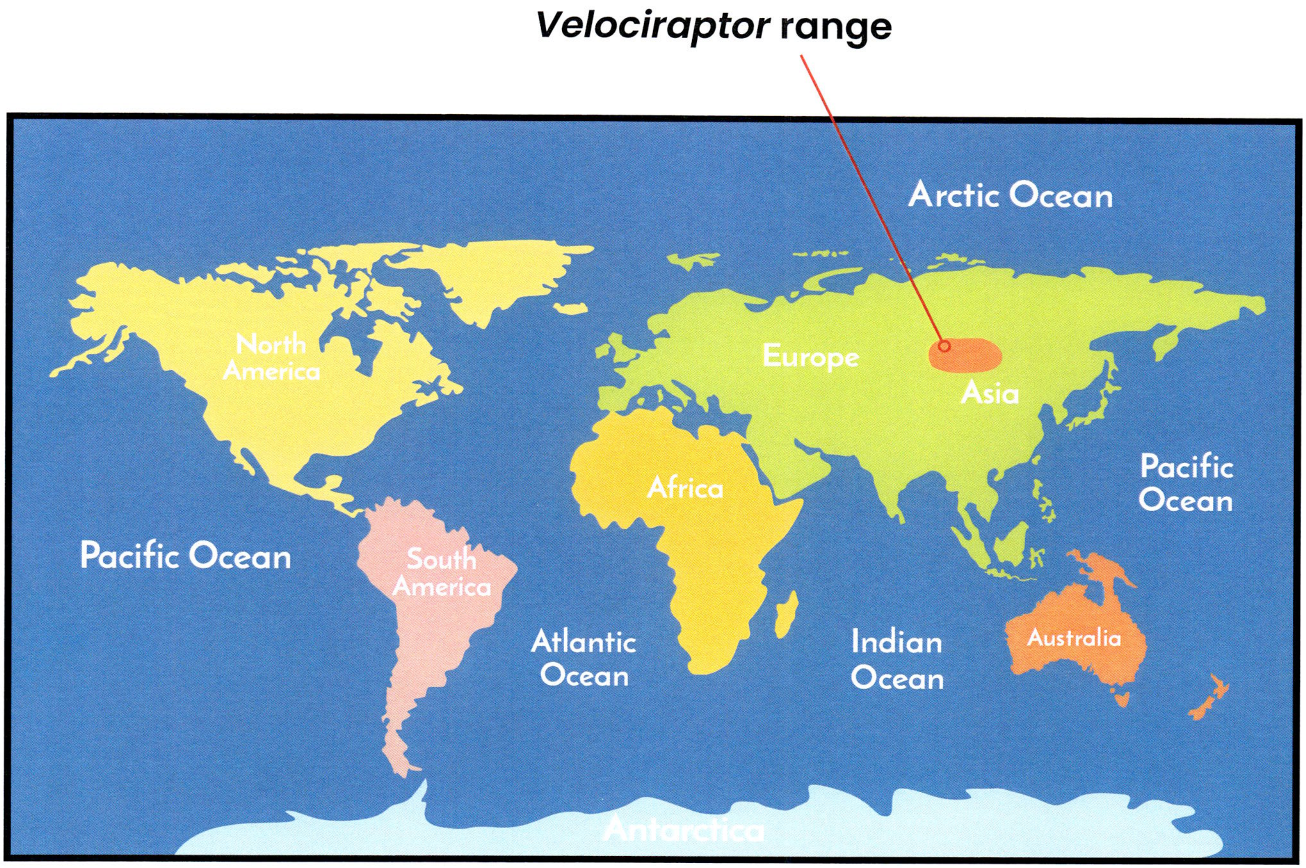

Velociraptor *fossils have been found in China and Mongolia.*

Velociraptor was a small dinosaur.

It was about the size of a turkey.

Velociraptor was a **predator**.
It used its claws to hunt.

Velociraptor was a **carnivore**.

It ate small dinosaurs and **mammals**.

Velociraptor *hunted, but it probably also ate animals that were already dead.*

Velociraptor was covered in feathers, but it could not fly.

Did *Velociraptor* use its feathers to **attract** mates?

Velociraptor’s *feathers might also have helped it stay warm.*

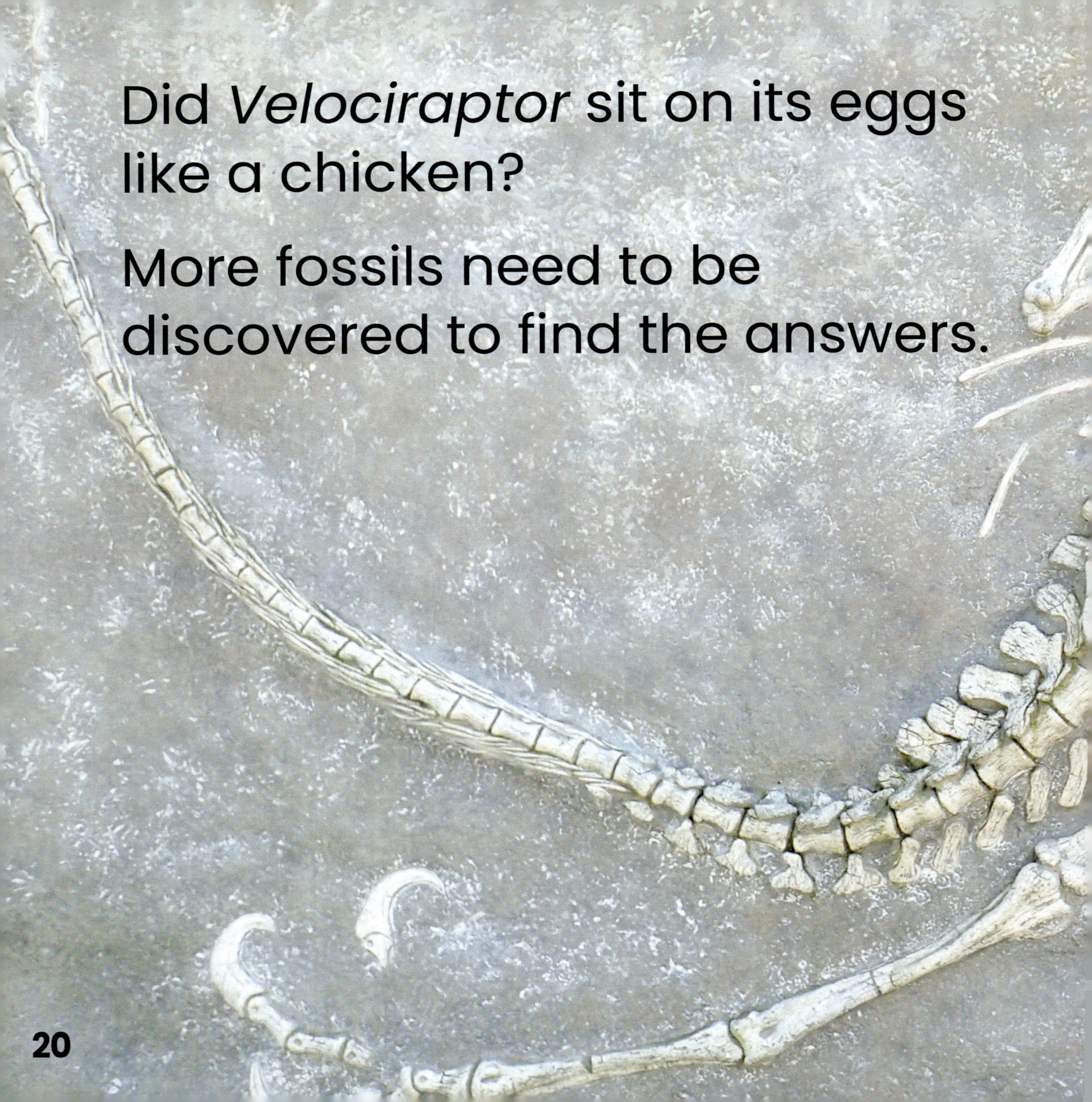

Did *Velociraptor* sit on its eggs like a chicken?

More fossils need to be discovered to find the answers.

Glossary

attract (UH-trakt): To cause interest or attention

carnivore (KAR-nih-vor): An animal that only eats meat

fossil (FAWS-uhl): The traces, prints, or remains of plants and animals that lived long ago

mammal (MAM-uhl): A warm-blooded animal that has hair or fur, and that feeds milk to its young

paleontologist (PAY-lee-en-TAW-luh-jist): A scientist who studies fossils to learn about past life on Earth

predator (PREH-duh-ter): An animal that hunts and eats other animals

talon (TA-luhn): A sharp, curved claw some birds use to catch animals for food

Index

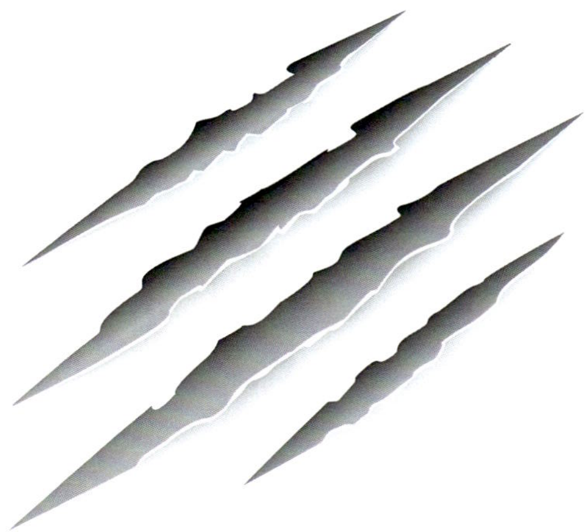

About the Author

Tracy Vonder Brink loves to learn about science and nature. She has written many nonfiction books for kids and is a contributing editor for three children's science magazines. Tracy lives in Cincinnati, Ohio, with her husband, two daughters, and two rescue dogs. She would like to touch a *Velociraptor* feather.

Written by: Tracy Vonder Brink
Illustrated by: Riley Stark
Designed by: Rhea Wallace
Series Development: James Earley
Proofreader: Melissa Boyce
Educational Consultant: Marie Lemke M.Ed.

Photographs:
Shutterstock: Arpad Benedek: p. 5; Rimma R: p. 7; Noiel: p. 13; Natalia Van D: p. 20-21

Crabtree Publishing

crabtreebooks.com 800-387-7650

Printed in the U.S.A./022024/PP20240115

Published in Canada
Crabtree Publishing
616 Welland Ave.
St. Catharines, Ontario
L2M 5V6

Published in the United States
Crabtree Publishing
347 Fifth Ave
Suite 1402-145
New York, NY 10016

Library and Archives Canada Cataloguing in Publication
Available at Library and Archives Canada

Library of Congress Cataloging-in-Publication Data
Available at the Library of Congress

Hardcover: 978-1-0396-9648-8
Paperback: 978-1-0396-9755-3
Ebook (pdf): 978-1-0396-9969-4
Epub: 978-1-0396-9862-8